the INTERNET of MYSTERIOUS THINGS

BY LISA SEACAT DELUCA

ART BY ADAM RECORD

ISBN-13: 978-0-9914152-2-9

Printed in China

First printing, March 2017

First Edition

Tap a NFC-enabled device to the hidden mysterious creatures for extras!

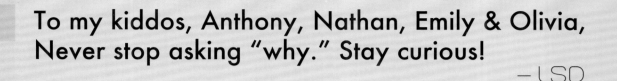

To my kiddos, Anthony, Nathan, Emily & Olivia,
Never stop asking "why." Stay curious!

— LSD

For Haven, Liam, Noah & Jill. Love you all!

—AR

The humans don't know, it is probably better that way.
They are glued to their devices all night and all day.

Smart homes, Bluetooth, and the Internet of Things,
everything is connected these days... so it seems.

Whenever a button is pressed in an app,
a mysterious creature will awake from its nap.

The creature will teleport to the location
and perform the task without hesitation.

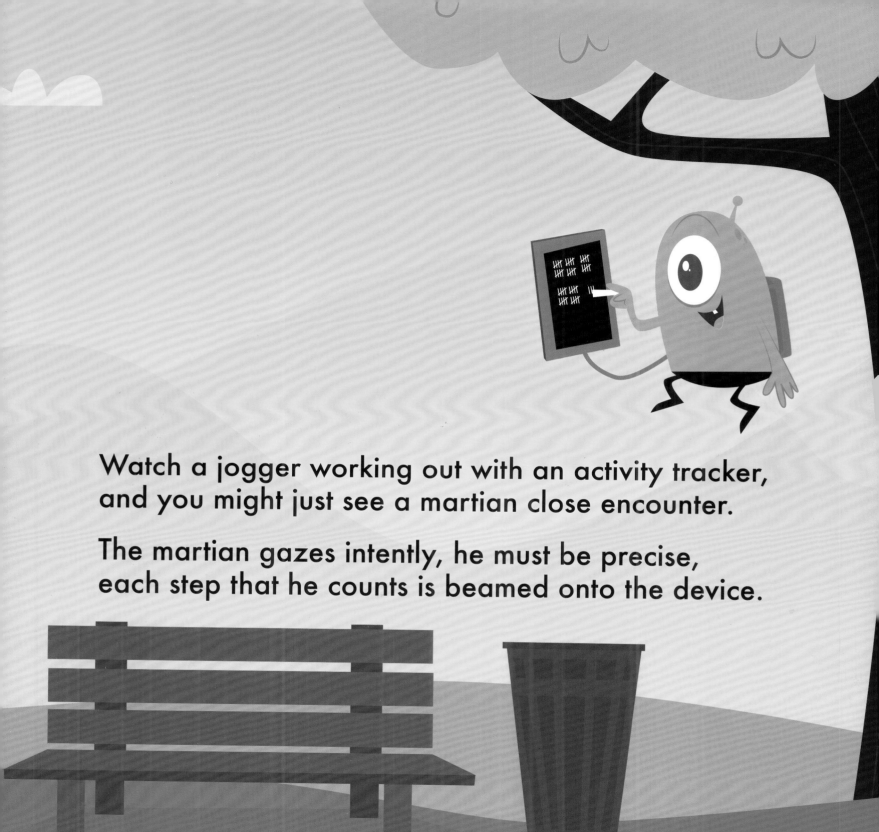

Watch a jogger working out with an activity tracker,
and you might just see a martian close encounter.

The martian gazes intently, he must be precise,
each step that he counts is beamed onto the device.

Listen carefully to a security alarm horn,
and you could hear the hooves of a unicorn.

He monitors for movement, doors opening, and beeps,
making humans feel safer, but he never sleeps.

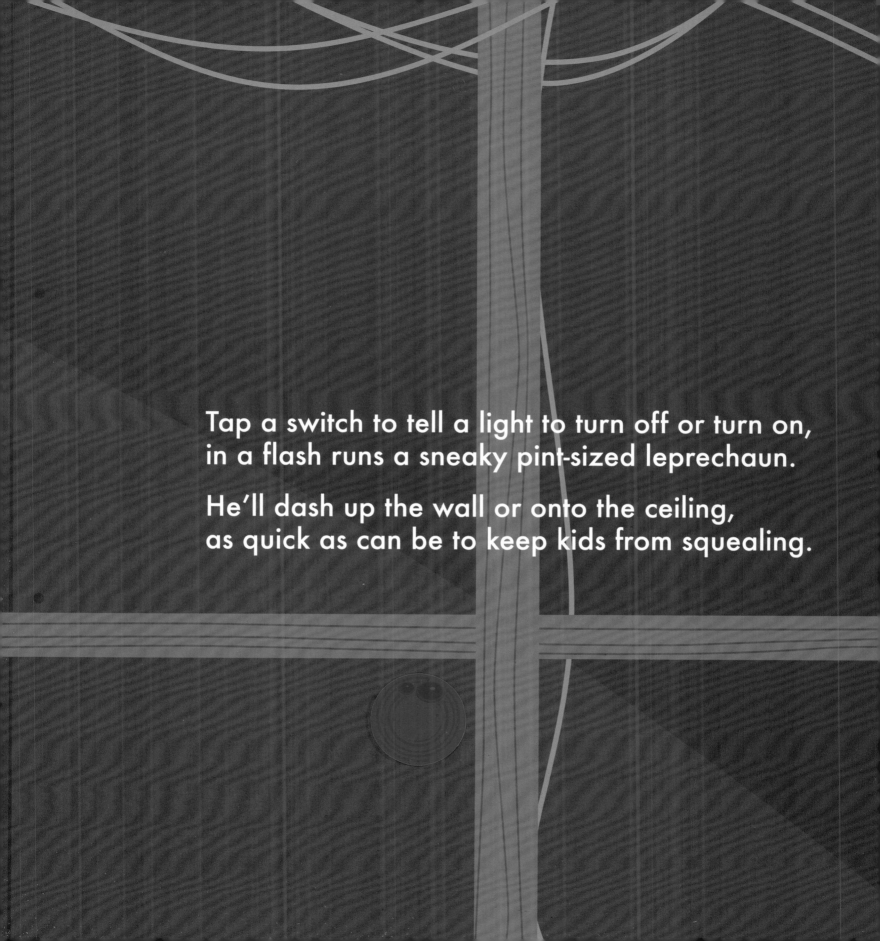

Tap a switch to tell a light to turn off or turn on,
in a flash runs a sneaky pint-sized leprechaun.

He'll dash up the wall or onto the ceiling,
as quick as can be to keep kids from squealing.

Open a door, without a traditional key,
is done by a zombie, not RFID!

When you are near, the door he'll unlock,
so his kids won't awake with the sound of a knock.

Find a location using GPS, when you are ready,
summons the help of at least one friendly Yeti.

Known as world-travelers, they give paths close inspection,
providing suggestions for street map directions.

Receive a notification, you're running low on eggs, it was sent by a monster with long hairy legs.

Inside your fridge, observing what you devour, taking note to remind you to purchase more flour.

Sip coffee from a Keurig that brews while you dream, a colorful dragon was what made it steam.

The dragon breathes fire to warm up the water, making the coffee flow hotter and hotter.

Split a dinner between friends, make sure it's paid,
the bill is divided by a singing mermaid.

She knows the order and calculates its amount,
transferring coins from each person's bank account.

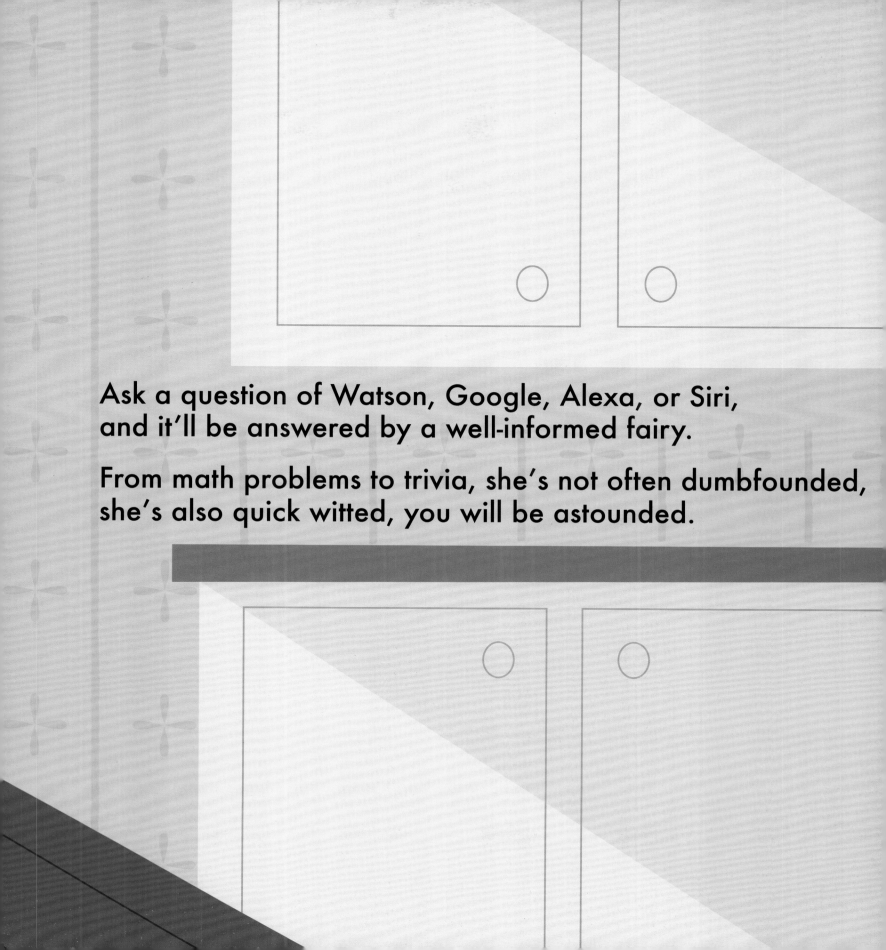

Ask a question of Watson, Google, Alexa, or Siri, and it'll be answered by a well-informed fairy.

From math problems to trivia, she's not often dumbfounded, she's also quick witted, you will be astounded.

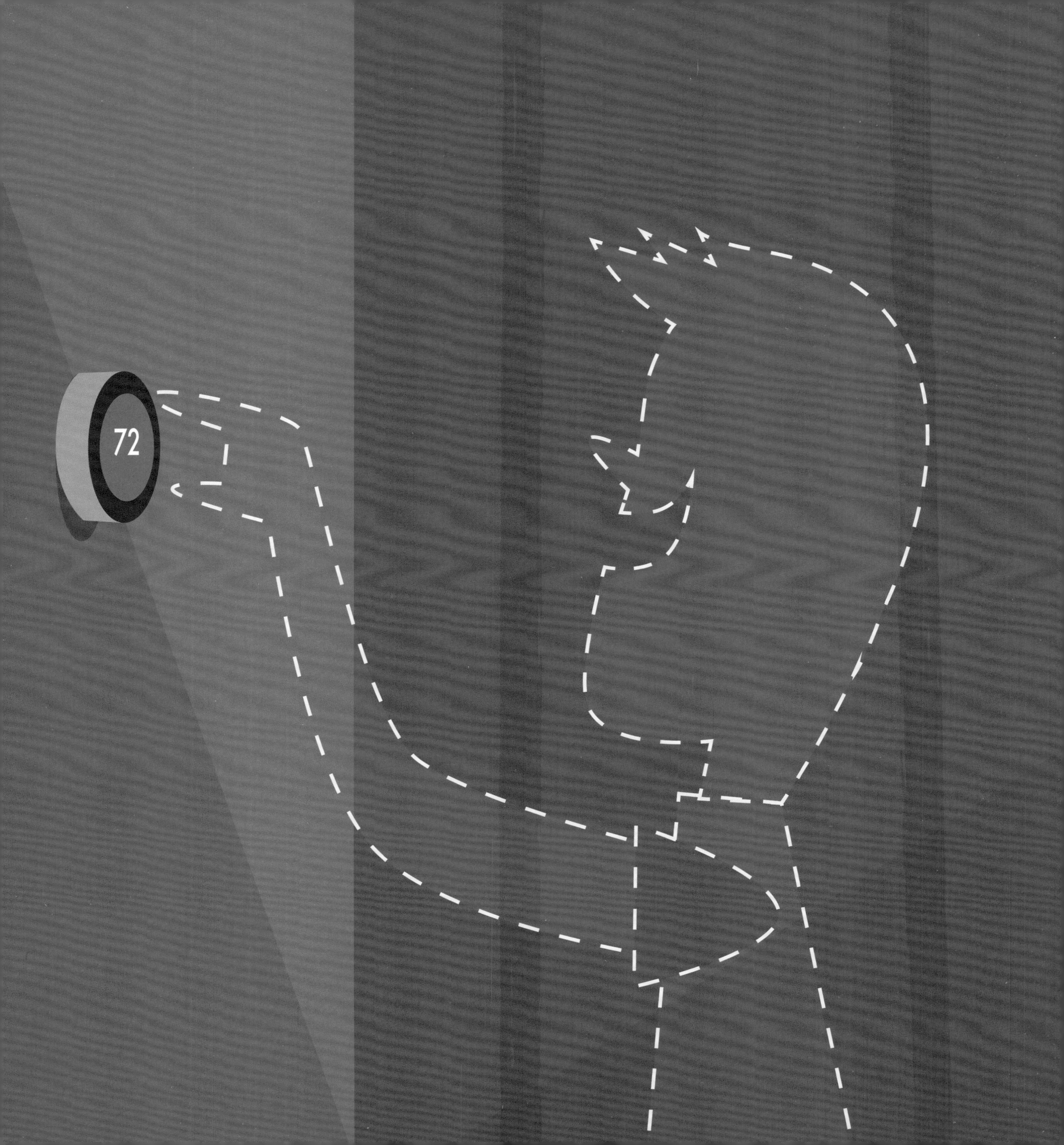
72

Adjust the temperature, turn up the fan,
and out pops a very invisible man.

To be truly invisible, he can't wear a suit,
but he'll warm your house with a respectful salute.

Future invention, what creature will it be?
I can't tell you right now, we must wait and see.

So put down your tablet, look up from your phone,
mysterious creatures are all around us,
we're never alone.

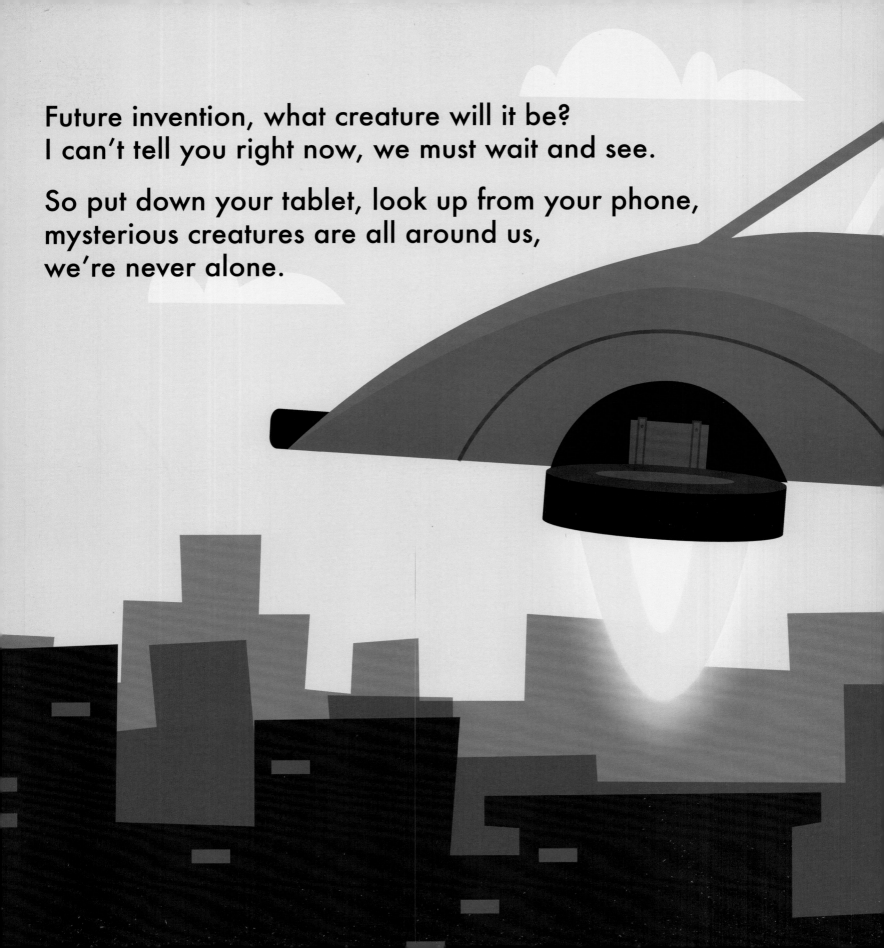

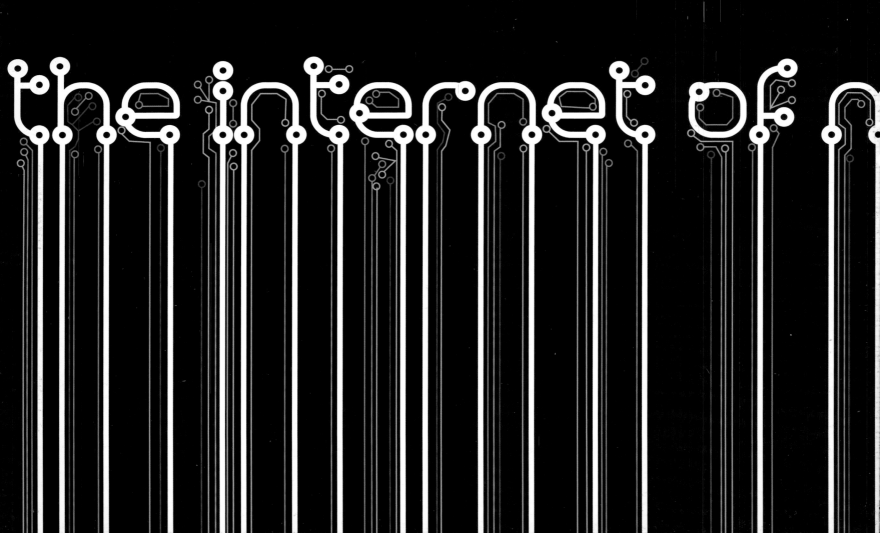